NOTES TO PSYCHE

Also by John Mack

*A Land Between Worlds: The Shifting Poetry
of the Great American Landscape*

Notes to Selfie: Bits of Truth in a Phoney World

At Their Home: Marseille

Revealing Mexico (co-author)

John Mack

NOTES TO PSYCHE

Mindful Thoughts for Thought-Filled Minds

LIFE CALLING

ISBN 979-8-9857987-1-5

Life Calling
www.life-calling.org

Cover Design: Zana Moraes
Layout and Production: Duane Stapp

For my siblings. (They are also my friends.)

"Chess holds its master in its own bonds,
shackling the mind and brain so that the inner freedom
of the very strongest must suffer."

ALBERT EINSTEIN

Contents

Introduction
1

Mind/ful
5

Now/here
13

Re/action
21

Id/ology
29

Con/science
37

Know/ledge
45

Pre/sense
53

In/dependence
61

Non/material
69

Pre/occupation
77

Im/perfection
85

En/lighten
93

Spiritual/Life
101

Un/seen
109

Here/after
117

You/topia
125

About John Mack
134

About Life Calling
135

Introduction

The war within is the only war. Upon the battlefield of the mind wages the fight for the human Soul. Involvement in external conflict is but a distraction from this fact. As in any struggle that concerns one's freedom, there is an enemy and a savior. As *potential*, you are both.

For the savior to prevail, your task is twofold: first, you must *see* the battlefield in its entirety; second, you must *act* upon that realization. By contrast, to be unaware of the battlefield's borders is to be unaware of the battle. Take note, psyche: to fight blindly is to *believe* you are the savior while *playing* the enemy's game.

Welcome to the *inner universe*.

Insofar as our human struggles play out on Earth—a spherical accumulation of stardust—so too does the battle for the human Soul play out on the sphere of the mind's contents—an accumulation of thought. The question is: Are you *thought-filled*—pegged by gravity to earthly consciousness? Or are you *mindful*—having ascended to the space of higher awareness?

The inner battle, like a game of chess, is a matter of positioning. Unlike a game of chess, the savior is positioned off the chessboard entirely. To be beyond the black-and-white grid of the Earth's game is to be off-planet: no us versus them, no winner or loser, no taking sides. Opponents on a chessboard are divided and, therefore, conquered—no matter who wins. *Ah*…can you sense the higher enemy in that last sentence? Can you sense the higher savior? Indeed, the war within is a battle of a higher realm. What content do you identify with? Against what content do you fight? If you are serious about this, let's be clear: ascension begins where "you" ends.

As in the magic of a stereogram, whereby a shift of the eyes reveals hidden content, so too must you perform an inner shift in the mind to reveal *its* content. Can you discern where one field stops and an-*other* field begins? Go ahead, look around. Can you shift your positioning in consciousness to see the battlefield in its entirety? Remember: to ascend in dimensions is to reveal a lower landscape.

What is needed is a short-circuiting of the mind, and the notes in this book are tools toward such outcome. Like the Japanese *koan* or the *kireji* in traditional haiku, these are not answers but thoughts styled to challenge the limitations of thought itself. However, it must be acknowledged that, like any spiritual practice (meditation, yoga, prayer, breathwork…what have you), tools are always limited; they may bring you to the edge of the battlefield, but that leap is yours to take.

Make no mistake, psyche, this is a battle for the Heavens. The enemy wants your mind full. The savior *is* mindful. Mind you, this concerns us all. These are "Notes to Psyche."

NOTES TO PSYCHE

The hello and the goodbye—one in the gesture of a wave.　　　5

Beyond form, beyond image—love joins two pieces
of separate puzzles.

We must question ourselves without doubting ourselves.
The former preserves vitality, the latter kills it.

Life is an inquiry—not an expectation of a question as prostituted by an answer.

To want something out of life is like squeezing
an orange; it reduces the full fruit down to
sweet pleasure—life's zest tossed in the waste bin.

To be "bored out of one's mind" is either an impossibility or a blessing. As if the clear sky were causing gray weather, to be "out" of that which is producing the experience is an impossibility. On the other hand, to be so gray as to have fully had enough of it would be a blessing, for one would get "out" of the source of the experience.

"*Personne*" in French means "nobody."

To take things personally is to be at war with oneself.
To reject that battle is the seed of war.

Now here = nowhere. 13

Zero—neither positive nor negative—is the space that holds all numbers. Its shape complete, its value empty—zero is two in one.

Like the yin-yang in the modern *taijitu*, time does not
exist outside the clock's circle.

16 Try your best to catch that flight, but don't be late to wherever you end up.

It is often said spirituality is a process akin to peeling the layers of an onion. But is this undoing not the mind's addiction to process? If any part of the onion remains, so too does the "who you are not," which is mutually exclusive from "what you Are." The final layer cannot be a layer but rather that which remains of the onion itself. As the mind cannot undo itself, the goal becomes the process, rather than throwing away the onion from the get-go. The ending of the onion would be the realization that, other than the edible ones, there never was an onion to begin with.

17

18   Only in our minds have we ever not been where we are.

Follow your heart, but with one caveat: Emotions may *seem* of the heart, but they are mind-based—reactions to pleasures or pains. To truly follow one's heart is to navigate above both.

20 If we believe in a path, we're off it.

Hate is the highest apathy.

22   Once one knows what action is, all reactions are overreactions.

It is not life that throws us curveballs but rather our
assumption of what is straight.

The first barrier has always been unawareness. The second inevitably risks following the overcoming of the first: inaction.

Most everyone has realized a rounded scoop of ice cream brings a moment of sweetness to life. Few have realized a rounded scoop of life brings ice cream to a new level of sweetness.

In essence, nerve-wracking anxiety and joyous
excitement are two different responses to the same
illusion; both are reactions to what isn't.

26

Are we open-minded enough to consider we might
be utterly close-minded? (Perhaps we sit with that
before proclaiming "Yes.")

When one lands on the nonduality of love, then hate—
28   now having no opposite—receives the light of truth.
Hate was always the reaction to a lie.

Today's reigning ideology is, and always has been, ideology itself.

30 Belief is but a type of technology—a psychological life preserver.

The only difference between an individual's hatred and mass genocide is the degree of its manifestation.

The limitations of psychology as an authority on sanity reside in the horizontality of its sample set. If 99% of the human population is positioned in dimension A— the dimension of separation—and the remaining 1% in dimension B—the dimension of union—then what good is an assessment of sanity based on observations from every country and region, every age group and gender if drawn solely from dimension A? Is not the healthiest degree of insanity still insane? When it comes to dimensions of consciousness, a sample set drawn from a single dimension—no matter how diverse—is as misleading as a taxonomist claiming truths about a Kingdom through the study of a species. In matters of consciousness, the anomaly, although not the norm, may very well be the "normal."

The most infinitesimal degree of self-importance begins at "bigger than the Universe."

34 Life has a spiraling truth that seductive spiders are born to spin. Beware winged creature: there is no depth in flatness.

Belief is the worship of a thought.   35

The difference between obligation and responsibility is a matter of hierarchy. Obligation serves one's self-abandonment—an assumed-responsibility toward a horizontal (but pedestaled) other. Responsibility is direct alignment—the effortless delivery of a vertical order.

Intellect cannot see eternity, it can only think about it. Science is blind.

The highest limitation of science has been that it assumes its view is the only valid view. Unsurprisingly, many scientists would deny this. To be open-minded to new facts doesn't disprove being close-minded in empiricism.

What science calls "fact" is a subjective agreement.
That there is night and day is a fact, is it not? But this is
a fact *on*-planet, not off-planet. Higher positionings
change agreements.

Within the Kingdom of God, a kingdom called "science." In the kingdom of science, a churchyard. Upon churchyard (no church), a tombstone. Entombed, the Kingdom of God.

In the dance of expansion, awareness follows inquiry's lead. 41

The scientist in awe of science—an experience of thought.
The scientist in awe of nature—an experience of beauty.

42   A scientist need not 'see' nature to be a scientist in awe,
for nature inspires much thought. One day, science might
open to the thought of including nature in its awe.

Cold hard facts are cold and hard. We must soften our
hearts with the warmth of a new Fact altogether.

44   At some point, science will have to transcend the
unknown to the unknowable.

Knowledge is not an answer but an exclamation revealed when not asking.

You can only be something when you're something.
But you can never be everything until you're nothing.

To be centered is to not have one.

Pay attention to the timing, for even in the warmth of water the sweetest green teas turn bitter.

Resentment rears its head in the resistance to one's deeper wish to forgive. Fortunately, the wish to forgive rears *its* head in the resistance to facing the totality of the pain.

If we do something when there is something to do, why, then, when there is nothing to do, do we resist the nothing there is to do? It is right *here*—in this impossibility of doing—wherein lies the transfer of power to the effortless.

There is a big difference between having an opinion and being opinionated. Where *you* have the former, the latter has *you*.

Might the only thing missing be a habitual appreciation
for what's not missing?

If there were ever such thing as a human savior, it would
be the children. Today's diplomas are but confirmations
of the crucifixion.

Boredom is a matter of life or death. Imagine lying on your deathbed and gazing for the last time at the world around you. Go ahead, give it one final look—make it *real*. Now ask yourself: Was there ever such thing as boredom, or was it merely a forgetting of life's brevity?

Boredom, at its core, is the absence of gratitude.

Mind-altering substances, no matter how glorified, are the failed vaccines of the familiarity disease.

"Falling in love" is commonly used as a mechanism to deliver the other to where we Are not.

The selfie is a phoney identity.

No matter how much haste might feel like catching up,
it is always late.

Presence isn't a place in time, but a positioning out
of time. Only when out of time do we have all the time
in the world.

Embracing one's 'imperfections'—as is so often
recommended—does not free oneself from judgment
but merely pardons the judge.

The question is not whether there is freewill, but rather, whether there isn't. It should be noted, however, that such question is not intended for the one who asks, but rather for the one to whom it is posed.

Hate is freedomless. That is where hate comes from—the hating to not be free: the choice as simple as the equation.

The true rebel is free and is therefore not a rebel.

Human relationships are fields full of obstacles.
To dissolve an obstacle is to roam the field more freely.
Translation: Human relationships are fields full
of opportunities.

Self-responsibility is—*at once*—an act of freedom and an act of love, for all three are one and the same.

The individuality of the single bee is essential to
the community of the hive—the purer each buzz, the
sweeter the honey.

68 The peace beyond like and dislike is the freedom
of preference.

The wish to belong brings the urge to possess.
Longing for belonging, we want *it* to belong to us.

70 Consumerism is to fill the shopping bag with that one thing not found.

Yearning for the sacred is yearning for an image.
The desire for a sacred relationship, experience, or object
is doomed to fail from the get-go. The true sacred
dwells beyond all imagery.

Emptiness is the blueprint of your temple.

The best thing we can do for our self-esteem is to not need one.

The difference between an individual belief and full-blown religion resides in the number of believers.

The last illusion is in the label.

If reason cannot prove God, then it cannot disprove God. God proves Itself, and, in so doing, shatters reason.

When it comes to Higher education, the student must
test all teachers in order to graduate.

78   If ignorance were bliss, there would be no suffering.

Love is an embrace without arms.

To be preoccupied is exactly that: the occupying of your vacancy before you do.

Fear is always an anticipation—the reaction to an illusion; it can only arise when face to face with what isn't.

The joy of life (for an adult) is as the joy of play for a child; growing up means accepting yourself as your sole authority.

Either way you look at it, the invitation "to be there or be square" is an invitation to squareness. To *show up* is to attend neither, whether you end up going or not.

True peace is not found in security, but rather where one never trusted security to be.

The difference between "I'm perfection" and "imperfection" resides in either the remembering or the forgetting of a certain inner space. (And, of course, in an apostrophe.)

86 If all gain and all loss serve inner growth, then there is
only gain.

Pain is the cry calling you back to your power.

If a gatekeeper stood in front of a door covered in an abundance of locks, how many different keys could they sell you before you pushed to check if the door was ever locked to begin with? A lifetime's worth? Assuming the door was never locked, who was the gatekeeper all along?

The wonder of a broken heart is that no piece is lost.

The difference between feeling safe *with* someone and
finding safety *in* someone is the difference between
intimacy and smothering it.

There's a maturity in childhood that most adults have left behind.

We act "imperfectly" when we forget we are perfect.

Just as the moon reveals the light within the darkness,
so too, our inner light burns for self-reflection.

Unlike most adults, a child is 100% mature, for immaturity is the disparity between age and its respective responsibility.

As children, we're told we can be anything when we grow up—a firefighter, an astronaut, a ballet dancer, the president. When we reach adulthood there are only two things we can be: asleep or awake.

"Unconditional love for my child," you say? And what if it were not your child? (Realizing conditions is like a game of hide-and-seek.)

As the purity of sunlight in the vastness of space, so too,
your light shines most pure in the vastness of you.

To care more about enlightenment than care itself is a backward equation.

Where one believes there is light, there is only darkness.
Where one knows there is darkness, there is only light.

Part of the problem is we were told we're not enough.
The entirety of the problem is we continue to buy into it.

Spiritual practice (when a gilded tool) glimmers as the golden hammer...for want of a nail, the Kingdom is lost.

Like altitude, *ascension* is a distancing from surface.
Like space, *depth* is a distance from gravity.

Spiritual practice, if necessary, need not be anything more than a simple reminder. Spiritual seeking, if practiced, need not be anything less than a lifelong attempt.

104   Empathy relates. Sympathy hijacks. Compassion empowers.

When all belief ends, spirituality goes with it, for all that is left is Life as it Is.

Spirit is always a kindred one.

Spiritual practice is practice for life. But since such practice occurs *within* life, it separates. What runner practices during the race? Practice for life must *be* life. Where living eliminates practice, practice eliminates living.

Perhaps real spiritual inquiry is to put down all the books, opt-out of all the talks, cancel the meditation sessions, toss your *Notes to Psyche*, and sit with what's right there.

If love is unconditional, then, unlike good-versus-bad or right-versus-wrong, it can have no equal opposite. Fear, its closest "opposite," is a dimension unto itself, albeit one that sets conditions.

One cannot be more loving than they already are.

110

The issue is that we Are not.

To make the point "Subjectivity is consciousness" is but a Euclidian one, *i.e.*, a primitive notion. In fact, it is the point itself that makes such claim, attempting to defend the size and shape of its nonexistence.

To live with resentment—whatever the degree—is to sacrifice one's quality of life. Certainly, nobody acting as an authority over their own state of being would choose that for themselves. To resent is to worship.

Devotion is not about picking and choosing what is worthy of one's devotion—that is devotion to judgment. The highest devotion is giving one's entirety to nothing in particular.

The spark of honesty begets order, no matter the perceived chaos of that transition.

The eternal gift is the 'you' that keeps opening.

The Hereafter: *here, after* a realization.

The Afterlife: *after life* as one has always known it to be.   117

118   To un-know is to arrive at the unknown.

It's just as it reads: To want something *out of life* is to want something dead.

The entrance to paradise operates like a sign that reads

"Beware of Dog," where, unbeknownst to us, there is no dog, and we are the ones barking.

We don't run toward virtuality (digital worlds, drugs, experiences, belief systems) to escape reality, but rather to escape the virtual Hell we have projected over Paradise.

Before the ego's death, we think we know. After its death, we know we don't.

To die is to have your cake. To live is to eat it too. 123

There is a big difference between falling *from* Paradise and falling *for* Paradise. Whereas the former is natural, the latter is artificial.

Utopia and dystopia are two opposing sides of the same coin: the coin of systemhood. The true youtopia is but a realization—one which, inevitably, upgrades the quality of systems.

About 99.9% of the journey is learning who we are not.
The remaining 0.1% is a 100% different journey altogether.

We strive to be exactly where we are, but that's not doable as we are already here. The difficulty of joy is in the ease.

Suffering is the upholding of identity; grieving, the witness of its dying; joy, the elation at its death.

To reflect upon a mirror's reflection is self-reflection.

Without reflection, there is only wall.

130 One act of honesty—in an instant—realigns the entire Universe.

The sweetest of humble pies is the one where it's realized no slice is missing.

Is it a tree, or do we call It a "tree"? The wonderful thing about trees is not that they are trees per se but that they might not be. What is your name?

About John Mack

John Mack is an author, artist, thought leader, and founder of the nonprofit, Life Calling. For years he has communicated his concern over the effects on humanity in the wake of the rapid proliferation of smart devices. Mack is the creator of Life Calling's signature program, *A Species Between Worlds: Our Nature, Our Screens*—an educational forum and interactive visual art experience exploring humanity's current migration from the natural world to the virtual world. He was an honoree of The Explorer's Club 50: "Fifty people changing the world who the world needs to know about." In addition, Mack serves as a board member of Fairplay, an organization that strives to create a world where kids can be kids, free from the false promises of marketers and the manipulations of Big Tech. He is a frequent speaker at conferences and universities on subjects of technology, awareness, and consciousness.

About Life Calling

Founded by John Mack in 2021, Life Calling is a non-profit organization dedicated to preserving our humanity in the Digital Age. Life Calling envisions a future where we remain grounded in our humanity—no matter the technological environment—ensuring we thrive in all that makes us human individually, collectively, and with nature. The organization assists in answering the call for expanded self-awareness. This is achieved through educational programming rooted in immersive experiences across the arts, culture, and nature, inclusive of publications and self-learning tools—each fostering and catalyzing a more balanced, nuanced and thoughtful perspective as we navigate the Digital Age.

www.life-calling.org

Printed in Great Britain
by Amazon